MathStart®

DOUBLING NUMBERS

Double the Ducks

by Stuart J. Murphy

illustrated by Valeria Petrone

HarperCollinsPublishers

LEVEL
1

10-03
17.00

To Kole—who, with his big sister, Samantha,
makes for double the fun
S.J.M.

To Double Trouble (Tommaso and Sara)
V.P.

The publisher and author would like to thank teachers Patricia Chase, Phyllis Goldman, and
Patrick Hopfensperger for their help in making the math in MathStart just right for kids.

HarperCollins®, ☂®, and MathStart® are registered trademarks of HarperCollins Publishers.
For more information about the MathStart series, write to HarperCollins Children's Books, 1350
Avenue of the Americas, New York, NY 10019, or visit our website at www.mathstartbooks.com.

Bugs incorporated in the MathStart series design were painted by Jon Buller.

Library of Congress Cataloging-in-Publication Data
Murphy, Stuart J.
    Double the ducks / by Stuart J. Murphy.
        p.   cm. — (MathStart)
    "Level 1."
    "Doubling numbers."
    ISBN 0-06-028922-8 — ISBN 0-06-028923-6 (lib. bdg.) — ISBN 0-06-446249-8 (pbk.)
    1. Multiplication—Juvenile literature.   [1. Multiplication.]   I. Title.   II. Series.
QA115 .M8654   2003   2001024321
513.2'13—dc21

Typography by Elynn Cohen   1   2   3   4   5   6   7   8   9   10   ❖   First Edition

Double the Ducks

I'm one busy person,
as you can see,
with a flock of five little ducks.

4

1

5

It takes lots of work.
There's just one of me
to care for my five little ducks.

6

I have only two hands
and so much to do
to look after my five little ducks.

9

I give them their food.

I bring three sacks a day

to feed to my

five little ducks.

With four bundles of hay
I make cozy nests
to shelter my five little ducks.

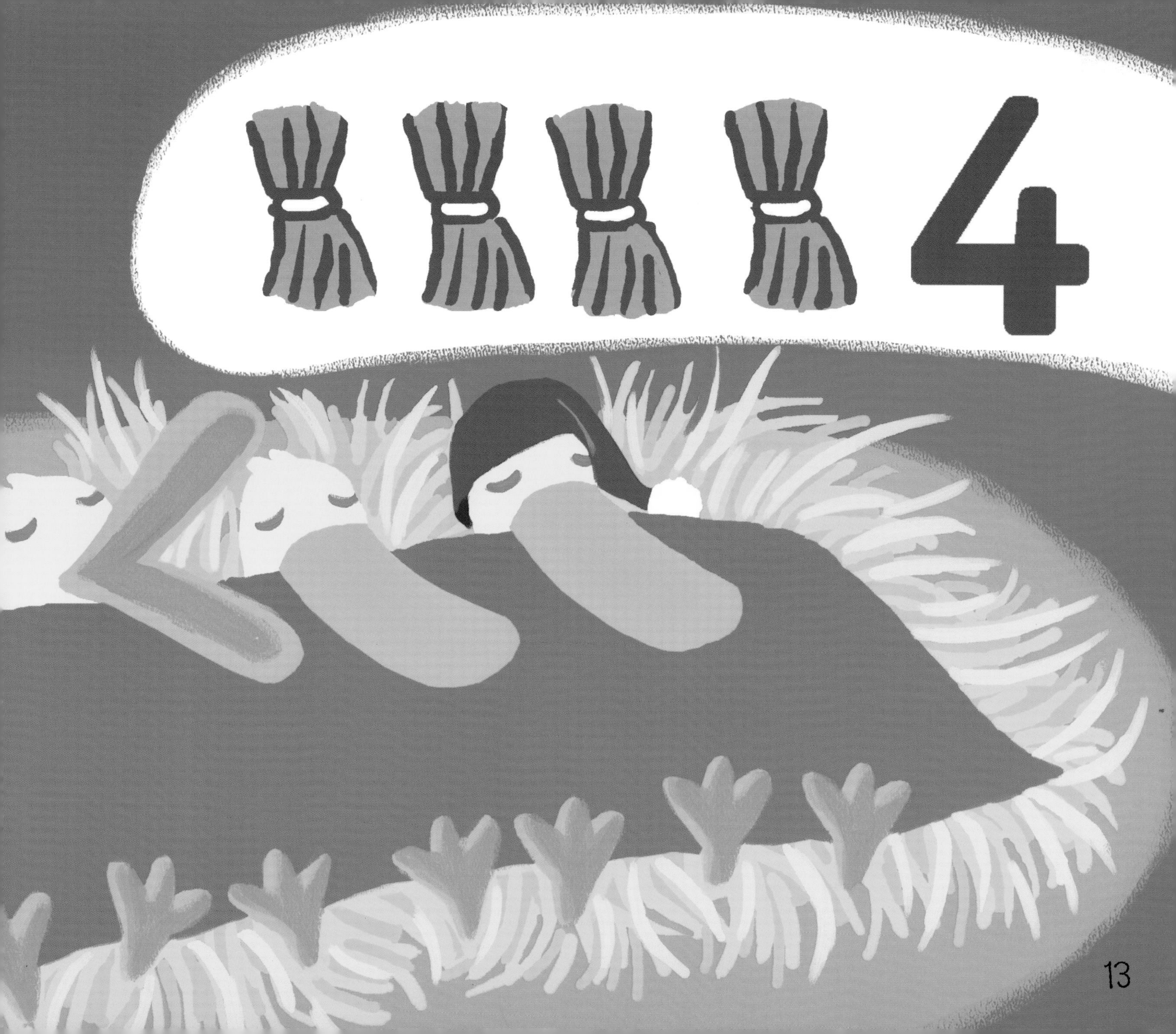
4

My ducks splash and slide.

They honk, quack, and glide.

It's fun to have five little ducks.

17

Now I have double the ducks
and it's double the work
to take care of my ten little ducks.

18

10

I need double the hay,
eight bundles in all,
to make nests for my ten little ducks.

20

8

I need double the food,
six sacks a day,
to feed to my ten little ducks.

23

I need double the hands
(that's four hands in all!)
to take care of my ten little ducks.

24

To have double the hands
I would need two of me
to look after my
ten little ducks.

26

2

It looks like I need a friend too.

31

In *Double the Ducks*, the math concept is doubling numbers. Adding a number to itself—for example, 3+3—is a beginning step in mastering basic addition.

If you would like to have more fun with the math concepts presented in *Double the Ducks*, here are a few suggestions:

• Read the story with your child and encourage him or her to count the objects (such as 2 hands or 3 sacks of food) mentioned in the story.

• Reread the story together with small objects at hand, such as buttons, marbles, or blocks. Encourage your child to use the objects to act out the story.

• Build a block tower between 1 and 10 blocks tall. Ask your child to count the number of blocks in the tower and then help your child build a second tower using the same number of blocks. Have your child find the total number of blocks it took to build both towers. Repeat this activity with different numbers of blocks.

• Using a set of dominoes, have your child pick out all the doubles.

• Help your child make a doubles book. Give each page a heading, such as "Double 1," "Double 2." Together with your child, think about objects that would go in each category. For example, "Double 1" might be a face with 2 eyes; "Double 2," a domino with 2 dots on each side; and "Double 3," an insect with 3 legs on each side.

Following are some activities that will help you extend the concepts presented in *Double the Ducks* into a child's everyday life:

Doubles Game: All players start with 10 points. On each turn a player rolls 2 dice. If doubles are rolled, the player receives the total number of points shown on the dice. If no doubles are rolled, the player loses 1 point. The first person to get 20 or more points is the winner.

Kitchen Calculation: When making a simple treat, like instant pudding, help your child double the recipe.

Mind-Reading: Tell your child that you thought of a number and then doubled it. Tell him or her the doubled number and ask your child to guess the original number. (For example, if the doubled number is 10, the correct answer would be 5.) If your child has difficulty, give him or her a group of small objects (buttons, paper clips, or the like) that total your doubled number and then separate them into 2 equal groups.

The following books include some of the same concepts that are presented in *Double the Ducks*:

- HOW MANY FEET IN THE BED? by Diane Johnston Hamm

- TWO OF EVERYTHING by Lily Toy Hong

- THE TOKEN GIFT by Hugh William McKibbon